UNLEASH THE POWER OF AZURE FILES FOR YOUR BUSINESS

I0490869

AZURE FILES
THE COMPLETE GUIDE

MASWOOD KHAN

CONTENTS

01

INTRODUCTION TO AZURE FILES

Azure Files is a cloud-based storage solution provided by Microsoft Azure, designed to provide secure, scalable, and highly available file shares for cloud-based and on-premises applications. With Azure Files, you can create and manage fully managed file shares that can be accessed from anywhere using standard SMB and NFS protocols.

Azure Files is ideal for organizations that need to store and share files between different applications, users, and devices, without the need to manage complex file servers or storage systems. Azure Files provides a highly available and scalable storage solution that can be accessed from anywhere, making it ideal for organizations with a geographically dispersed workforce.

Azure Files supports a range of storage tiers, including hot, cool, and archive tiers, allowing you to optimize your storage costs based on your data access patterns. Additionally, Azure Files provides a range of advanced features, such as Azure File Sync, disaster recovery, and SMB Multichannel, making it a powerful and flexible solution for a wide range of use cases.

In this guide, we'll explore the basics of Azure Files, including how to get started, how to manage your files, how to integrate Azure Files with other services, and best practices for using Azure Files effectively. We'll also cover some advanced features and troubleshooting tips to help you get the most out of your Azure Files deployment.

02

GETTING STARTED WITH AZURE FILES

Before you can start using Azure Files, you'll need to create an Azure account and subscription. Once you've done that, you can create a storage account and set up Azure Files. Here's how to get started:

Create an Azure account and subscription: If you don't already have an Azure account, go to the Azure website and sign up for a free account. You'll need to provide some basic information and a credit card number to create a subscription.

Create a storage account: Once you've created a subscription, you'll need to create a storage account to store your files. To do this, go to the Azure portal and click on "Create a resource." Then select "Storage account" and follow the prompts to create a new storage account.

Set up Azure Files: After you've created a storage account, you can set up Azure Files. To do this, go to the storage account and click on "File shares" under the "File service" section. Then click on "New file share" and follow the prompts to create a new file share.

Configure access to Azure Files: Once you've created a file share, you'll need to configure access to it. You can do this by creating an access key, which allows you to connect to the file share using the SMB or NFS protocol.

Upload and manage files: After you've set up Azure Files, you can start uploading and managing files. You can do this using a variety of tools, including the Azure portal, Azure Storage Explorer, or command-line tools like AzCopy.

By following these steps, you can quickly and easily set up Azure Files and start using it to store and share files. In the next chapter, we'll explore how to manage your Azure Files, including how to upload and download files, monitor storage usage, and configure backup and recovery options.

03

MANAGING AZURE FILES

Once you've set up Azure Files, you'll need to manage your files and ensure that your storage stays organized and secure. In this chapter, we'll explore some best practices for managing Azure Files, including how to upload and download files, monitor storage usage, and configure backup and recovery options.

Uploading and downloading files: To upload files to Azure Files, you can use tools like the Azure portal, Azure Storage Explorer, or AzCopy. You can also use APIs or SDKs to automate file uploads. To download files from Azure Files, you can use the same tools or download files directly from the file share using SMB or NFS.

Monitoring storage usage: It's important to monitor your storage usage to ensure that you don't exceed your storage limits or incur unexpected costs. You can use the Azure portal or Azure Storage Explorer to monitor your storage usage and set alerts for when your storage usage reaches a certain threshold.

Configuring backup and recovery: Azure Files provides built-in backup and recovery options, including point-in-time recovery and geo-replication. You can configure these options using the Azure portal or Azure Storage Explorer to ensure that your files are protected against data loss.

Managing access and security: To ensure that your files are secure, you'll need to manage access to Azure Files. You can do this by configuring access keys, setting up role-based access control (RBAC), or using Azure Active Directory (Azure AD) to manage authentication and authorization.

Configuring advanced features: Azure Files provides a range of advanced features, including Azure File Sync, SMB Multichannel, and Azure Private Link. These features can help you optimize your storage and improve performance. You can configure these features using the Azure portal or Azure Storage Explorer.

By following these best practices for managing Azure Files, you can ensure that your files are secure, organized, and easily accessible. In the next chapter, we'll explore how to integrate Azure Files with other services, including Azure Virtual Machines, Azure Kubernetes Service (AKS), and Azure App Service.

04

INTEGRATING AZURE FILES WITH OTHER SERVICES

One of the main benefits of using Azure Files is that it can be easily integrated with other Azure services, allowing you to build more powerful and scalable solutions. In this chapter, we'll explore how to integrate Azure Files with other services, including Azure Virtual Machines, Azure Kubernetes Service (AKS), and Azure App Service.

Azure Virtual Machines: Azure Files can be used as a shared file system for Azure Virtual Machines, allowing you to share files between VMs without the need for a separate file server. To do this, you'll need to mount the Azure File share as a network drive on the VM.

Azure Kubernetes Service (AKS): Azure Files can be used as a shared file system for AKS, allowing you to share files between containers running in the same pod or across multiple pods. To do this, you'll need to create a Kubernetes volume using the Azure Files storage class.

Azure App Service: Azure Files can be used as a shared file system for Azure App Service, allowing you to share files between web apps or deploy files to the file share using continuous deployment. To do this, you'll need to mount the Azure File share as a network drive on the web app.

Other services: Azure Files can also be integrated with other Azure services, such as Azure Batch, Azure Functions, and Azure Logic Apps.

These services can use Azure Files to store and share data, or to trigger automated workflows based on file events.

By integrating Azure Files with other Azure services, you can create more powerful and scalable solutions that can easily adapt to changing business requirements. In the next chapter, we'll explore some advanced features of Azure Files, including Azure File Sync, disaster recovery, and SMB Multichannel.

05

BEST PRACTICES
FOR AZURE FILES

Azure Files is a powerful and flexible storage solution that can be used to store and share files across your organization. To ensure that your Azure Files implementation is secure, performant, and cost-effective, it's important to follow best practices when designing and managing your file shares. In this chapter, we'll explore some best practices for Azure Files.

Choose the right performance tier: Azure Files offers two performance tiers: Standard and Premium. Standard tier is suitable for general-purpose file shares, while Premium tier is recommended for high-performance workloads that require low latency and high IOPS.

Optimize file sharing: To ensure that your file shares are secure and easily accessible, it's important to use the right file sharing protocols and authentication methods. Azure Files supports both SMB and NFS file sharing protocols and allows you to use access keys, Azure AD, or RBAC to manage authentication and authorization.

Monitor and manage storage usage: To avoid unexpected costs or performance issues, it's important to monitor your storage usage and set alerts for when you reach certain thresholds. You can use Azure Monitor to monitor your storage usage and set up alerts, and use Azure Storage Explorer to monitor and manage your file shares.

Use encryption and access control: To ensure that your files are secure, it's important to use encryption and access control mechanisms. Azure Files supports both encryption at rest and in transit, and allows you to use access keys, Azure AD, or RBAC to manage access control.

Configure disaster recovery: To ensure that your files are protected against data loss or service outages, it's important to configure disaster recovery options. Azure Files provides built-in options for point-in-time recovery and geo-replication, allowing you to replicate your file shares across regions for high availability and disaster recovery.

By following these best practices, you can ensure that your Azure Files implementation is secure, performant, and cost-effective. In the next chapter, we'll explore some common use cases for Azure Files, including data sharing, application migration, and hybrid cloud scenarios.

06

ADVANCED FEATURES OF AZURE FILES

Azure Files is a powerful storage solution that offers many advanced features to meet the needs of modern enterprises. In this chapter, we'll explore some of the advanced features of Azure Files, including Azure File Sync, disaster recovery, and SMB Multichannel.

Azure File Sync: Azure File Sync allows you to synchronize files between on-premises servers and Azure Files, enabling hybrid scenarios where you can store some files on-premises and others in the cloud. With Azure File Sync, you can also tier files to Azure Blob Storage, which can help reduce storage costs.

Disaster recovery: Azure Files provides built-in disaster recovery options, including replication of file shares across regions for high availability and point-in-time recovery. You can also use Azure Backup to back up your file shares to an offsite location.

SMB Multichannel: SMB Multichannel is a feature of Azure Files that allows multiple TCP connections to be established between the client and the server, increasing performance and availability. SMB Multichannel is automatically enabled for Premium tier file shares.

Identity-based authentication: Azure Files supports identity-based authentication using Azure Active Directory (Azure AD), which allows you to manage access to your file shares using Azure AD groups and conditional access policies. With identity-based authentication, you can enforce stronger security policies and reduce the risk of credential theft.

Advanced data protection: Azure Files supports advanced data protection features, such as Azure Information Protection and Azure Rights Management, which allow you to classify, label, and protect your files based on their sensitivity. With these features, you can ensure that your files are protected against unauthorized access and comply with data protection regulations.

By leveraging these advanced features of Azure Files, you can create more powerful and secure storage solutions that can meet the needs of modern enterprises. In the next chapter, we'll explore some common use cases for Azure Files, including data sharing, application migration, and hybrid cloud scenarios.

07

TROUBLESHOOTING AZURE FILES

Like any complex system, Azure Files can experience issues from time to time. In this chapter, we'll explore some common issues that you may encounter when working with Azure Files and provide some troubleshooting tips to help you resolve them.

Authentication issues: One of the most common issues with Azure Files is authentication. If you're having trouble accessing your file shares, make sure that you have the correct access keys or Azure AD credentials. You should also check your firewall settings to make sure that traffic is allowed to and from your Azure Files instance.

Performance issues: If you're experiencing slow performance when accessing your file shares, there are several factors that could be causing the issue. Check the performance tier of your file shares and make sure that it's appropriate for your workload. You should also check the size and number of files in your file shares, as well as the network connectivity between your client and the Azure Files instance.

Storage account issues: If you're having issues with your storage account, make sure that it's provisioned correctly and that you have enough storage capacity. You should also check for any account-level limits that may be causing issues.

File locking issues: If you're experiencing file locking issues, make sure that your applications are using the correct file locking mechanism. You should also check for any network connectivity issues that may be causing file locking issues.

DNS issues: If you're having trouble accessing your Azure Files instance, check your DNS settings to make sure that they're configured correctly. You should also check for any network connectivity issues that may be causing DNS resolution issues.

By following these troubleshooting tips, you can quickly identify and resolve common issues with Azure Files. If you're still having trouble, you can always reach out to Microsoft support for assistance.

08

CONCLUSION AND NEXT STEPS

In this book, we've explored the many features and capabilities of Azure Files, a powerful storage solution that can meet the needs of modern enterprises. We've covered everything from the basics of getting started with Azure Files to more advanced features like Azure File Sync, disaster recovery, and SMB Multichannel.

We've also discussed best practices for Azure Files, including how to manage and integrate Azure Files with other services, and how to troubleshoot common issues that you may encounter.

As you move forward with Azure Files, there are several next steps that you can take to further expand your knowledge and skills. These include:

Exploring additional Azure services: Azure Files is just one of many services available on the Azure platform. Consider exploring other services like Azure Blob Storage, Azure Data Lake Storage, and Azure Disk Storage to learn more about the different storage options available in Azure.

Learning more about hybrid cloud scenarios: Azure Files is a great solution for hybrid cloud scenarios, where you need to store some files on-premises and others in the cloud. Consider learning more about how to set up and manage hybrid cloud scenarios with Azure.

Studying for Azure certification: If you're interested in building your skills in Azure, consider studying for an Azure certification.

Azure offers several certifications for different skill levels and job roles, including the Microsoft Certified: Azure Administrator Associate and the Microsoft Certified: Azure Solutions Architect Expert.

By continuing to learn and explore Azure Files, you can

develop the skills and knowledge needed to create powerful and effective storage solutions for your organization.

09

APPENDIX: GLOSSARY OF AZURE FILES TERMINOLOGY

Here are some common terms and acronyms related to Azure Files that you may encounter:

Azure Files: A fully managed file share service in Azure.

File share: A logical unit of storage in Azure Files that contains files and folders.

Storage account: A container for data objects in Azure, including file shares in Azure Files.

Access key: A security credential used to authenticate with an Azure storage account.

Azure Active Directory (Azure AD): Microsoft's cloud-based identity and access management service.

SMB (Server Message Block): A protocol used for file sharing and communication between devices on a network.

NFS (Network File System): A protocol used for file sharing between Linux and Unix systems.

Azure File Sync: A service that allows you to sync files between on-premises servers and Azure Files.

Disaster recovery: The process of recovering from a disaster that results in data loss or system downtime.

SMB Multichannel: A feature that allows SMB to use multiple network connections to improve file transfer performance.

Azure Site Recovery: A service that helps you protect your data and applications by orchestrating replication and failover between Azure regions or on-premises sites.

Azure Monitor: A service that provides real-time monitoring and insights into the performance and health of Azure resources.

By understanding these terms and acronyms, you can better navigate and use Azure Files to its full potential.